The Muscle Book

Paul Blakey

Bibliotek Books

Dedication

The idea for this book germinated when I started teaching anatomy and physiology to dance students at the Merseyside Dance Centre, Liverpool, England.

It was obvious from my attempts to find suitable textbooks that most were inappropriate. What was needed was something like a car repair manual. You know the sort - it starts with a diagram of the car so you can locate the bit you want, followed by detailed information about specific parts showing how they relate to the whole.

That's what I've tried to do for you. I hope you like it.

Many thanks to Gaynor Owen, John and Mary Salisbury, and Nicky, Gemma and Saul for helping me.

And to all my teachers everywhere I say a huge thank you.

Paul Blakey

Disclaimer

The author and publisher of this material are not liable or responsible to any person for any damage caused or alleged to be caused directly or indirectly by the information in this book. If you are in doubt, consult a physician.

Published in 1992 by Bibliotek Books
ISBN 1 873017 00 6
Reprinted 1992, 1993, 1994, 1995, 1996, 1997
This revised edition published 1998,
reprinted 1999, 2000, 2001-5

Contents

How to use this book

This is your personal body parts manual. It is about muscles - what they look like, what they are attached to, what they move, what happens when they weaken, what to watch out for to avoid injury, and some simple finger first-aid.

Apart from a few pages at the beginning and at the end, all the muscle pages have a similar format. They include most or all of the elements shown below.

The **name** of the muscle, with an explanation of the origins of the term.

A **line drawing** of the muscle

The **body language of weakness** - a description of movement difficulties which indicate problems involving the muscle.

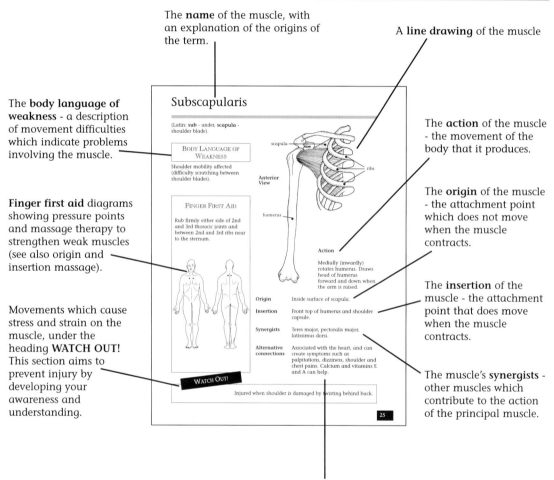

The **action** of the muscle - the movement of the body that it produces.

Finger first aid diagrams showing pressure points and massage therapy to strengthen weak muscles (see also origin and insertion massage).

The **origin** of the muscle - the attachment point which does not move when the muscle contracts.

Movements which cause stress and strain on the muscle, under the heading **WATCH OUT!** This section aims to prevent injury by developing your awareness and understanding.

The **insertion** of the muscle - the attachment point that does move when the muscle contracts.

The muscle's **synergists** - other muscles which contribute to the action of the principal muscle.

Subscapularis

(Latin: **sub** - under, **scapula** - shoulder blade).

BODY LANGUAGE OF WEAKNESS

Shoulder mobility affected (difficulty scratching between shoulder blades).

FINGER FIRST AID

Rub firmly either side of 2nd and 3rd thoracic joints and between 2nd and 3rd ribs near to the sternum.

scapula

ribs

Anterior View

humerus

Action

Medially (inwardly) rotates humerus. Draws head of humerus forward and down when the arm is raised.

Origin	Inside surface of scapula.
Insertion	Front top of humerus and shoulder capsule.
Synergists	Teres major, pectoralis major, latissimus dorsi.
Alternative connections	Associated with the heart, and can create symptoms such as palpitations, dizziness, shoulder and chest pains. Calcium and vitamins E and A can help.

WATCH OUT!

Injured when shoulder is damaged by twisting behind back.

25

Sometimes I include additional information which you may wish to consider in relation to specific body problems - for example about nutrition. This is found under the Alternative Connections heading.

On pages 6, 7, 8, and 9 you will find front and back views of the body showing the skeleton and muscles. These are reference pages which help you to see where muscles are located in relation to the whole body.

You can refer back to these from individual muscle pages, or use them as a kind of index if you know where a muscle is but not what it is called.

Anatomical terminology

I have tried to keep jargon to a minimum in this book, but some anatomical terminology is essential. This is simply because it is the most economical way of describing movement or relations of body parts to one another.

Regarding muscle actions, it is necessary to learn the following terms;

- **Flexion.** This means bending or folding. Flexor muscles generally curl the body forward into the foetal position. Toe flexors curl the toes.

- **Extension.** This means stretching out. Extensor muscles stretch the body, as in bending backwards with arms over the head.

Flexors and extensors oppose one another.

- **Rotation**. This means turning round, like a wheel. In anatomical terms it means movement around the axis of a bone.

- **Abduction.** This means taking or drawing away. In body terms it is a movement to the side by one muscle moving away from another, as in lifting a limb away from the body.

abduction

adduction

- **Adduction.** This means drawing inward. It is a sideways movement opposite to abduction, for example bringing a limb towards the centre of the body.

Adductors and abductors oppose one another.

Origin and Insertion Technique

You will come across this term frequently in the Finger First Aid sections. It is a method of stimulating or 'waking up' a weak muscle. The origin and insertion points are the places where the muscle attaches to the bone. All you have to do to stimulate the muscle is to place your fingers at these points and rub firmly across the fibres.

This technique works very well on muscles that have been strained or overworked. It stimulates the circulation and particularly the lymphatic system. It con be repeated frequently until the weakness and stiffness have gone.

Reference page 1

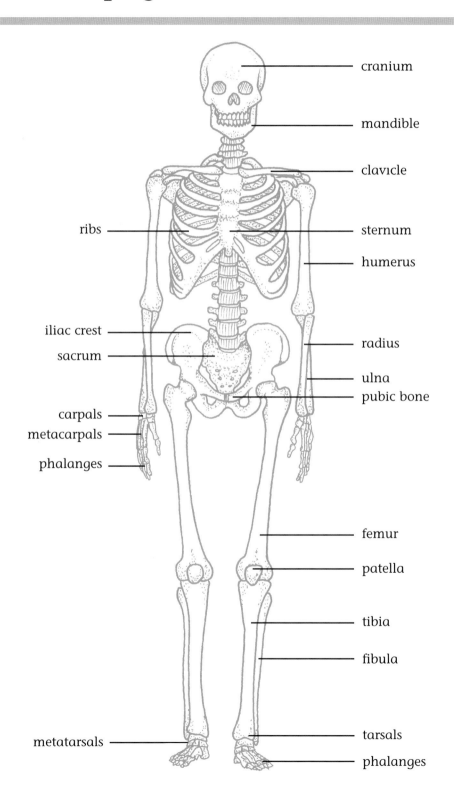

cranium

mandible

clavicle

ribs

sternum

humerus

iliac crest

sacrum

radius

ulna
pubic bone

carpals

metacarpals

phalanges

femur

patella

tibia

fibula

metatarsals

tarsals

phalanges

Reference page 2

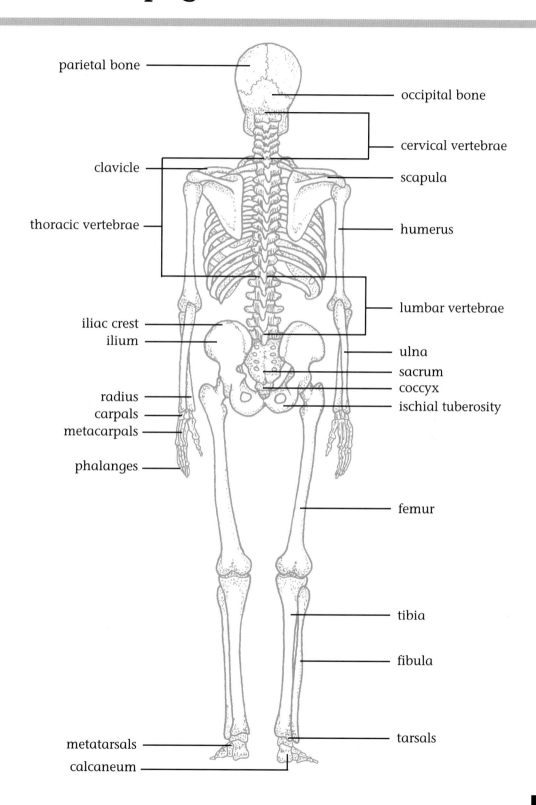

parietal bone

occipital bone

cervical vertebrae

clavicle

scapula

thoracic vertebrae

humerus

lumbar vertebrae

iliac crest

ilium

ulna

sacrum

coccyx

radius

ischial tuberosity

carpals

metacarpals

phalanges

femur

tibia

fibula

tarsals

metatarsals

calcaneum

Reference page 3

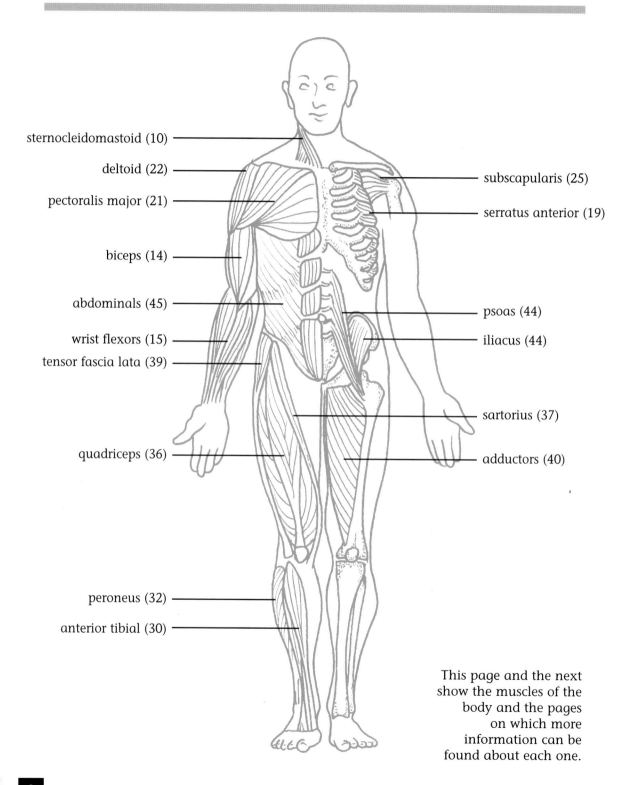

sternocleidomastoid (10)

deltoid (22)

pectoralis major (21)

biceps (14)

abdominals (45)

wrist flexors (15)

tensor fascia lata (39)

quadriceps (36)

peroneus (32)

anterior tibial (30)

subscapularis (25)

serratus anterior (19)

psoas (44)

iliacus (44)

sartorius (37)

adductors (40)

This page and the next show the muscles of the body and the pages on which more information can be found about each one.

Reference page 4

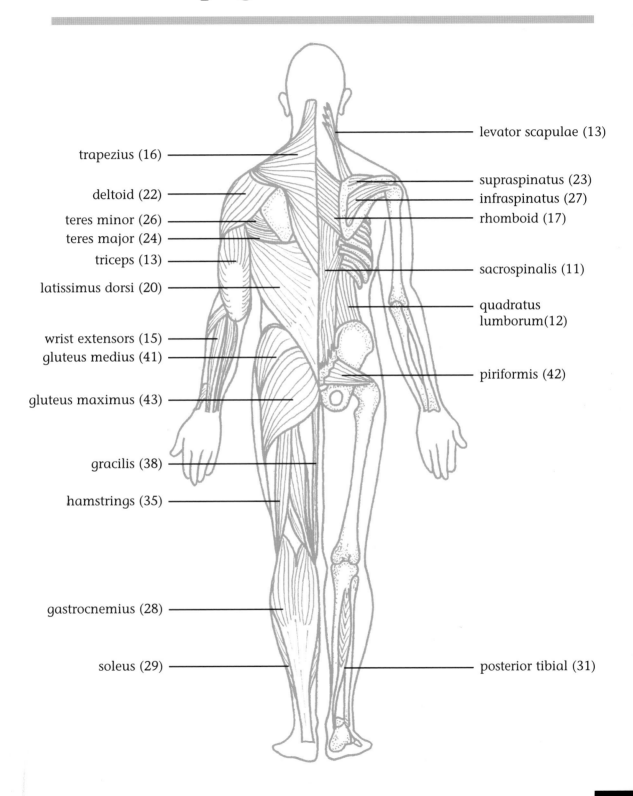

trapezius (16)

deltoid (22)

teres minor (26)

teres major (24)

triceps (13)

latissimus dorsi (20)

wrist extensors (15)

gluteus medius (41)

gluteus maximus (43)

gracilis (38)

hamstrings (35)

gastrocnemius (28)

soleus (29)

levator scapulae (13)

supraspinatus (23)

infraspinatus (27)

rhomboid (17)

sacrospinalis (11)

quadratus lumborum(12)

piriformis (42)

posterior tibial (31)

Sternocleidomastoid

Attached to the **sternum**, **clavicle** and **mastoid** bone.

BODY LANGUAGE OF WEAKNESS

Difficulty raising head from supine (lying on your back) position.

Head turns towards the weak side when lying down.

Neck extension increases, forming an 'S' curve.

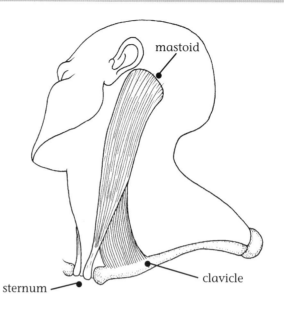

mastoid

clavicle

sternum

FINGER FIRST AID

To strengthen these muscles, rub the origin and insertion.

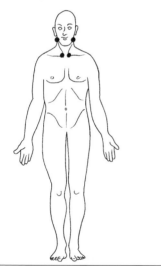

Action	Both together, flexes head. One only, rotates head and draws it towards shoulder.
Origin	Top of the sternum and first third of clavicle.
Insertion	Mastoid process of the skull, at the back of the head, behind the ears.
Synergists	Neck flexors.
Alternative connections	These muscles are also associated with the sinuses and can be affected by allergies, producing headaches and neck tension. B vitamins can help.

 **WATCH OUT!**

Violent whiplash movements cause injury to these muscles. They play a minor role in supporting the head in relation to balance reflexes and even minor strains can upset the body's structural alignment.

Sacrospinalis

(Latin: **sacrum** - holy, **spina** - thorn).

Weakness on one side will cause a sideways bending. Both sides weak will result in a round-shouldered, slouching posture. The strong side will often stand out like a rope or cord down one side of the spine.

FINGER FIRST AID

To strengthen, press firmly with your thumbs either side, down the length of the spine. Also press firmly (but not too hard) either side of the navel.

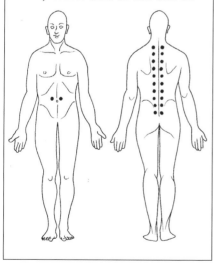

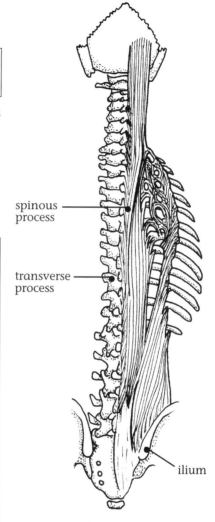

spinous process

transverse process

ilium

Action

Extension, lateral flexion and rotation of vertebral column; lateral movement of pelvis. (Bending backwards, bending sideways and twisting spine).

Origin

Separate slips of muscle arising from the sacrum, crest of the ilium, spinous processes, transverse processes and ribs.

Insertion

Into the ribs, transverse processes, spinous processes and occipital bone.

Alternative connections

Closely associated with the emotional aspects of bladder problems, (being unable to 'let go' of tension). Difficulty in eliminating the unwanted. Vitamins C and A can help.

WATCH OUT!

Avoid bending, stretching forward and lifting at the same time.

Quadratus Lumborum

(Latin: **quad** - four, **lumbar** - loin).

Weakness on one side will show as an elevation of the 12th rib and a curve in the lumbar vertebrae (leaning away from the weak side).

FINGER FIRST AID

To strengthen, massage firmly either side of L5 (fifth lumbar vertebra). Also just inside the top front edge of the iliac crest and the top half of the inside of the thigh.

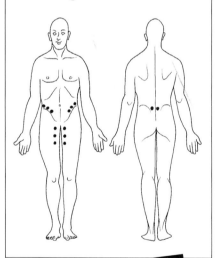

WATCH OUT!

Excessive side bending can injure this muscle, especially if done quickly.

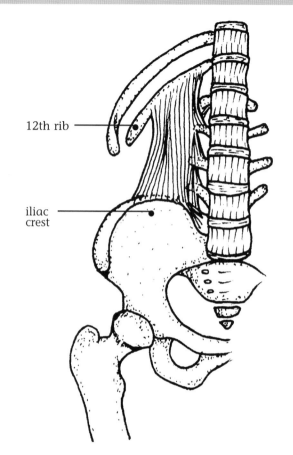

12th rib

iliac crest

Action	Lateral flexion (side bending) of lumbar vertebrae, depression of 12th rib; assistance of diaphragm in inspiration.
Origin	Top rear of iliac crest, and ilio-lumbar ligament.
Insertion	Bottom edge of 12th rib, transverse processes of the upper four lumbar vertebrae.
Synergists	Internal and external oblique abdominals, psoas major.
Alternative connections	Can often be involved when there is an intervertebral disk injury.

Triceps

(Latin: **tri** - three, **caput** - head).

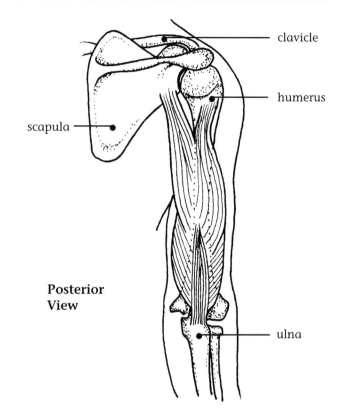

clavicle

humerus

scapula

Posterior
View

ulna

BODY LANGUAGE OF WEAKNESS

Arm remains bent if you try to straighten it above the head.

FINGER FIRST AID

Rub origins and insertion (top and bottom of humerus at the back).

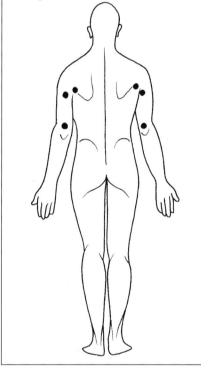

Action	Extends (straightens) forearm.
Origin	Top rear of humerus, outer edge of scapula below joint, rear of humerus just above the elbow.
Insertion	Ulna, below elbow.
Alternative connections	The proper function of this muscle can be affected by problems involving sugar metabolism. Vitamin A can help, as can avoiding excess refined sugar.

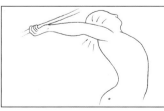

WATCH OUT!

Falling on the hand when the arm is bent or forceful throwing can injure the triceps tendon.

Biceps

(Latin: **bi** - two, **caput** - head).

BODY LANGUAGE OF WEAKNESS

When standing the elbow is held in extension (straight).

FINGER FIRST AID

Rub origins and insertions (top and bottom of humerus at the front).

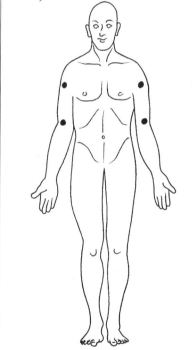

WATCH OUT!

Injury to the biceps tendon can occur when lifting or throwing heavy objects.

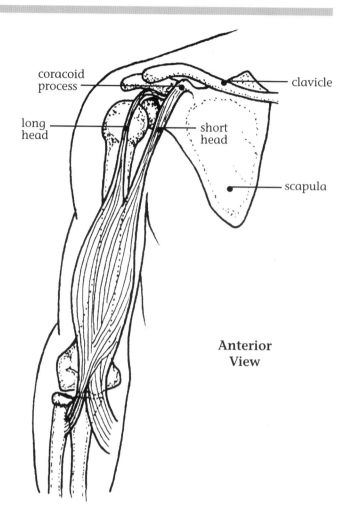

coracoid process — clavicle

long head — short head

scapula

Anterior View

Action	Flexes (bends) and supinates (turns palm upwards) the forearm. Also active when shoulder is flexed.
Origin	Short head: scapula, tip of coracoid process. Long head: top of the scapula, at the joint.
Insertion	Radius and deep fascia (connective tissue) of the forearm flexor muscles.
Alternative connections	Associated with nervous tension and insomnia.

Wrist Flexors/Wrist Extensors

Wrist benders and stretchers, and finger movers.

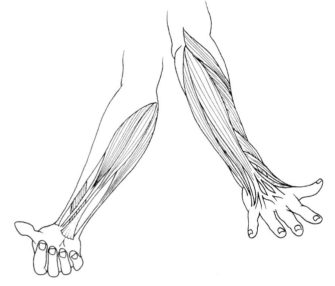

Inability to grip.

FINGER FIRST AID

Pain on the outside of the elbow can often be relieved by firm pressure into the origin (medial and lateral epicondyle of humerus) and belly of these muscles.

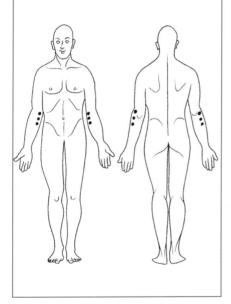

WATCH OUT!

Most injuries of the wrist involve falling onto the hand, either spraining or fracturing the joint.

Action	**Flexors** flex the wrist (palms upwards, bend wrist up towards body). **Extensors** extend the wrist (palms downwards, bend hand up towards body).
Origin	**Flexors:** medial epicondyle, (bump on the inside bottom of humerus), upper radius and ulna and interosseus membrane (membrane between the two bones). **Extensors:** lateral epicondyle, (bump on the outside bottom of humerus) and upper parts of the bones and interosseus membrane of the forearm.
Insertion	**Flexors:** carpal, metacarpal bones and phalanges. **Extensors:** carpal, metacarpal bones and distal phalanges (fingers).
Alternative connections	Tension in these muscles often means that you can't 'let go' of something in your life.

15

Trapezius (upper, middle and lower)

(Latin: **trapeza** - table, four sides not parallel).

FINGER FIRST AID

Midway between the shoulder and neck press firmly. Massage down either side of the spine from the neck to the bottom ribs. Pay special attention to the area between the shoulder blades and spine.

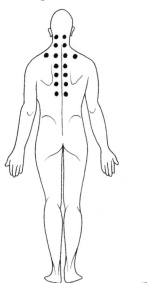

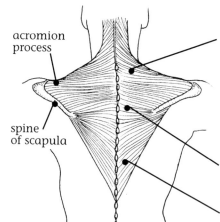

acromion process

spine of scapula

Action

Upper: adducts and rotates scapula, laterally flexes neck and head, rotates head away from the side which is contracting.
Middle: adducts and slightly elevates scapula.
Lower: rotates scapula, helps maintain spine in extension.

Origin	**Upper**: base of the skull, along spine to C7 (7th cervical joint). **Middle**: spinous processes of 1st to 5th thoracic vertebrae. **Lower**: spinous processes, 6th to 12th thoracic vertebrae.
Insertion	**Upper**: lateral third of clavicle and acromion process. **Middle**: upper border of spine of scapula. **Lower**: medial third of spine of scapula.
Synergists	**Upper**: Levator scapulae, sternocleidomastoid. **Middle**: rhomboids.
Alternative connections	The upper trapezius is associated with eye and ear problems. The middle and lower can be especially affected by infections, sore throats and fevers. Vitamins A, B and C can help.

WATCH OUT!

These muscles often become hypertonic (too tight) due to dance (particularly ballet) training, causing discomfort and restriction of mobility. Most neck and upper back stiffness in dancers is due to excess tension in this area.

Rhomboids

(Greek: **rhomb**: an equilateral parallelogram).

Complaints of aching and soreness between shoulder blades. Excessive military posture (shoulders pulled back), a sign of tension.

FINGER FIRST AID

Deep massage between shoulder blades and stretching, pressing the scapula away from the spine.

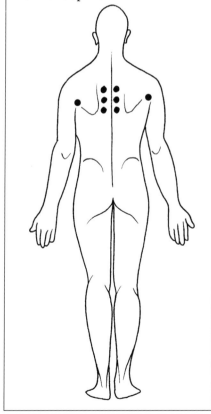

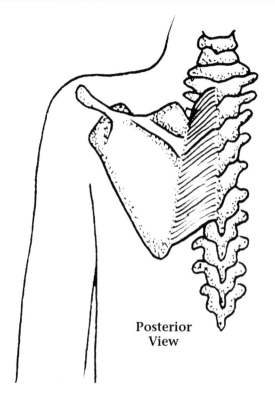

Posterior View

Action	Adducts (pulls towards spine) the scapula.
Origin	Spinous processes of the 7th cervical vertebra and the upper 5 thoracic vertebrae.
Insertion	Medial (inner) border of scapula.
Synergists	Trapezius, levator scapulae and latissimus dorsi.
Alternative connections	Associated with liver function.

WATCH OUT!

Like the trapezius, usually hypertonic (too tight).

17

Levator Scapulae

(Latin: **levator** - lifter, **scapula** - shoulder blade).

FINGER FIRST AID

Press firmly the belly of the teres minor muscle, also the intercostal space below the 1st rib near to the sternum (these are reflex points). Also massage origin and insertion (often the tendon of this muscle can be felt like a tight cord where it inserts into the scapula).

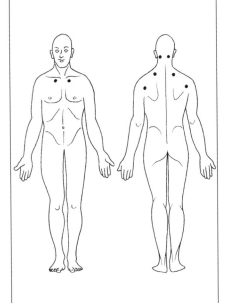

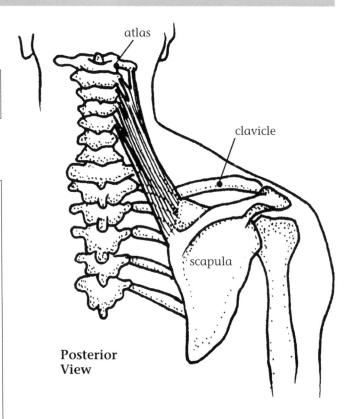

Posterior View

Action	In combination with trapezius, elevates and adducts scapula. When scapula fixed, flexes and slightly rotates cervical spine.
Origin	Transverse processes of four upper cervical vertebrae.
Insertion	Top of scapula closest to the spine.
Synergists	Rhomboids and trapezius.

WATCH OUT!

Usually hypertonic (too tight), affecting the mobility of the neck in rotation and side bending.

18

Serratus Anterior

(Latin: **serra** - saw, **anterior** - in front).

BODY LANGUAGE OF WEAKNESS

Difficulty pushing with arms straight out in front. Holding a weight in front will cause scapular 'winging'.

FINGER FIRST AID

Press firmly either side of the spine at the level of T3-T5 (3rd to 5th thoracic joints). Also between 3rd, 4th and 5th ribs near to the sternum.

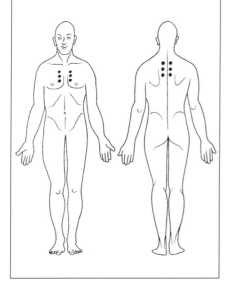

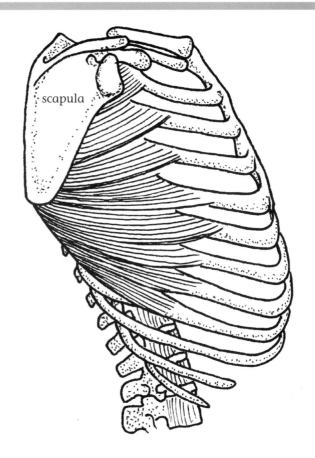

scapula

Action	Draws scapula forward against the ribs.
Origin	Outer surface of the upper 8 or 9 ribs.
Insertion	Inner surface of scapula along the medial edge nearest the spine.
Alternative connections	Can be affected by chest conditions or breathing difficulties. Vitamin C helps.

WATCH OUT!

Imbalances between serratus anterior and rhomboids or trapezius all cause tension problems in regard to shoulder mobility.

Latissimus Dorsi

(Latin: **latus** - broad, **dorsum** - back).

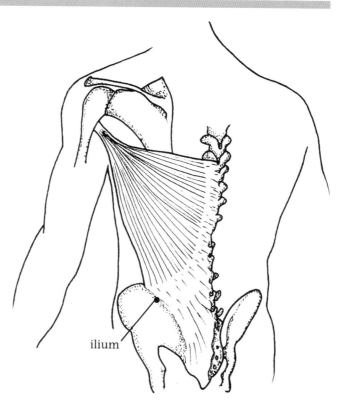

ilium

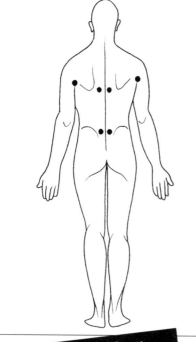

WATCH OUT!

Damage to this muscle can affect shoulder mobility.

Action	Extends, adducts and rotates the humerus medially (draws the arm back and inwards towards the body).
Origin	A broad sheet of tendon originating from the spinous processes of the six lower thoracic and the five lumbar vertebrae; also the posterior crest of the ilium.
Insertion	Twists to insert into the humerus just below the shoulder.
Alternative connections	Very sensitive to problems with sugar metabolism. Can also be an indication (if chronically weak), of possible allergies and intolerance to caffeine and tobacco. Avoid excess sugar. Vitamin A helps.

Pectoralis Major (Clavicular and Sternal)

(Latin: **pectoris** - the breast).

FINGER FIRST AID

These muscles are markedly affected by emotional stress and often a few minutes of light contact on the frontal eminences (points on the forehead, above the eyes), will do more than hours of massage. If the problem is simple physical exhaustion then rub the origin and insertion.

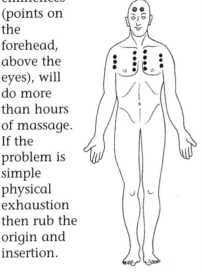

Anterior View

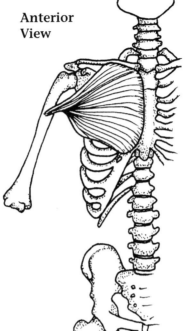

Action

Medial rotation of the humerus.
Clavicular: flexes the shoulder and horizontally adducts humerus.
Sternal: adducts humerus towards opposite iliac crest, major anterior shoulder stabiliser.

Origin

Clavicular: front surface of sternal half of the clavicle.
Sternal: sternum and adjacent parts of upper 7 ribs.

Insertion — **Clavicular**: top of the humerus under the front edge of the deltoid.
Sternal: same as clavicular.

Synergists — **Clavicular**: Biceps, sternal part of pectoralis major.
Sternal: Latissimus dorsi, subscapularis, teres major.

Alternative connections — This muscle is connected to reflexes which involve both the stomach and emotional centres in the brain. Possibility of problems with allergies. Vitamin B helps.

WATCH OUT!

The insertion of this muscle is the most likely area for injury, but usually this is limited to injuries sustained during weight training or other strength activities such as wrestling, shot putting or javelin throwing.

Deltoid

(Greek: **delta** - the fourth letter in the Greek alphabet, shaped like a triangle).

BODY LANGUAGE OF WEAKNESS

Difficulty holding arm out to the side.

FINGER FIRST AID

Front of body, between 3rd, 4th and 5th ribs near sternum. Back of body, either side of 3rd, 4th and 5th thoracic joints. Use firm pressure.

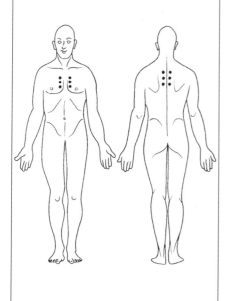

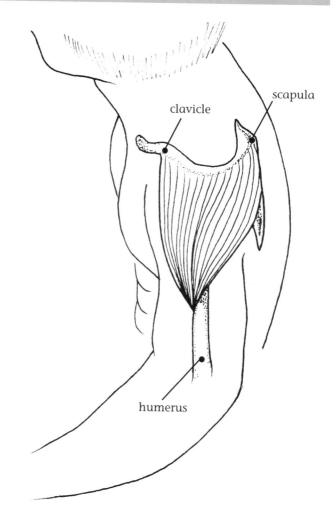

Action	Abducts the shoulder (lifts away from the body), flexes and extends the arm.
Origin	Clavicle and spine of the scapula.
Insertion	Side of humerus.
Synergists	Supraspinatus.
Alternative connections	Lung problems such as bronchitis, pleurisy, pneumonia and flu will affect this muscle. Vitamin C helps.

WATCH OUT!

A strong muscle, usually only injured if the shoulder joint is severely damaged.

22

Supraspinatus

(Latin: **supra** - above, spine of scapula).

Difficulty moving the arm away from the body.

Rub firmly the groove between the chest and arm. Also rub under the skull where it joins the neck.

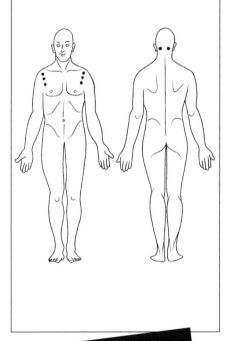

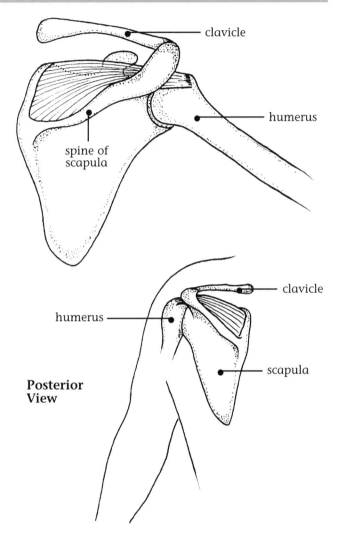

Posterior View

A shoulder dislocation will damage this muscle.

Action	Abducts arm with deltoid. Holds humerus in shoulder socket.
Origin	Scapula, above the spine of the scapula.
Insertion	Top of the humerus.
Synergists	Deltoid.
Alternative connections	Intense concentration, such as driving or working at a desk, often fatigues this muscle.

Teres Major

(Latin: **teres** - smooth, **major** - big).

Difficulty reaching behind and drawing the elbow backwards.

Rub origin and insertion. Also either side of 2nd and 3rd thoracic joints and between 2nd and 3rd ribs, 2-3 inches from the sternum.

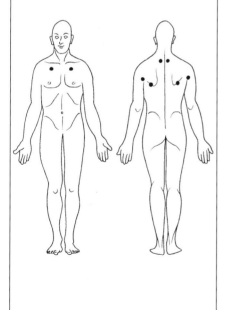

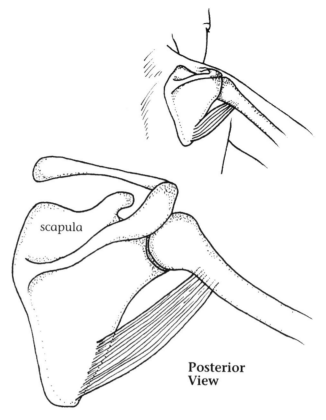

scapula

Posterior View

Action	Adducts (pulls inwards) and medially (inwardly) rotates humerus. Extends shoulder joint.
Origin	Bottom lateral edge of scapula.
Insertion	Back of humerus just below shoulder joint.

WATCH OUT!

Jerking the arm forwards sharply will damage this muscle.

Subscapularis

(Latin: **sub** - under, **scapula** - shoulder blade).

Shoulder mobility affected (difficulty scratching between shoulder blades).

FINGER FIRST AID

Rub firmly either side of 2nd and 3rd thoracic joints and between 2nd and 3rd ribs near to the sternum.

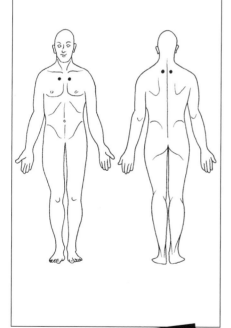

WATCH OUT!

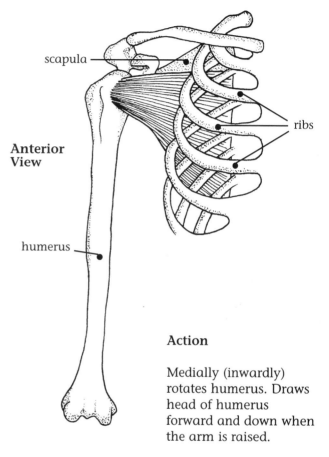

scapula

ribs

Anterior View

humerus

Action

Medially (inwardly) rotates humerus. Draws head of humerus forward and down when the arm is raised.

Origin	Inside surface of scapula.
Insertion	Front top of humerus and shoulder capsule.
Synergists	Teres major, pectoralis major, latissimus dorsi.
Alternative connections	Associated with the heart, and can create symptoms such as palpitations, dizziness, shoulder and chest pains. Calcium and vitamins E and A can help.

Injured when shoulder is damaged by twisting behind back.

Teres Minor

(Latin: **teres** - smooth, **minor** - small).

BODY LANGUAGE OF WEAKNESS

When the hands hang down by the side, the palms will face backwards.

FINGER FIRST AID

Rub firmly either side of 2nd and 3rd thoracic joints and between 2nd and 3rd ribs near to the sternum. Also rub origin and insertion.

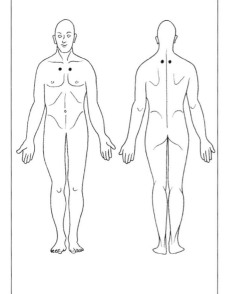

WATCH OUT!

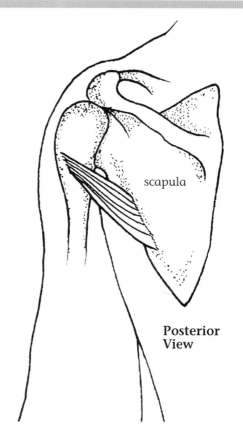

scapula

Posterior View

Action	Laterally (outwardly) rotates humerus. Also slightly adducts and extends. Stabilises humerus in shoulder socket.
Origin	Edge of scapula closest to arm.
Insertion	Top of humerus at the back. Shoulder capsule.
Synergists	Infraspinatus.
Alternative connections	Wrist and elbow problems can develop as compensation for a weak teres minor.

Jerking and twisting the arm will damage this muscle.

Infraspinatus

(Latin: **infra** - below, spine of the scapula).

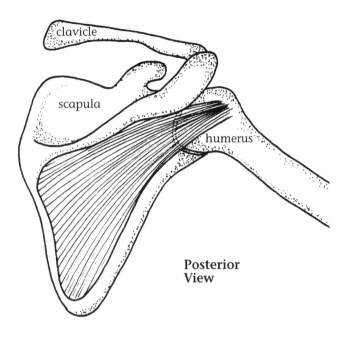

clavicle

scapula

humerus

Posterior View

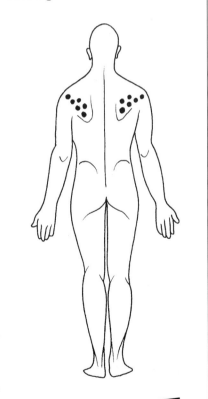

Action	Lateral rotation of humerus with teres minor. Stabilises head of humerus within the shoulder socket.
Origin	Middle two-thirds of scapula below spine of scapula.
Insertion	Top of humerus, shoulder joint capsule.
Synergists	Teres minor.

WATCH OUT!

One of the rotator cuff muscles. Always injured if the shoulder joint is damaged.

Gastrocnemius

(Greek: **gaster** - belly, **kneme** - leg).

BODY LANGUAGE OF WEAKNESS

Hyperextended knees.

FINGER FIRST AID

Stretch and massage whole muscle. Also press firmly 2 finger widths above and either side of the navel and either side of the 10th, 11th and 12th thoracic vertebrae.

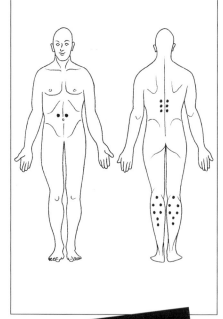

Action

Plantar flexion of foot (pointing the foot or rising onto demi-pointe).

Origin

Back of the leg, just above the knee.

Insertion

Via the Achilles tendon, into the heel bone.

Synergists

Soleus, posterior tibial, peroneus longus/brevis, flexor hallucis and digitorum longus.

Alternative connections

Emotional stress, fatigue and mild shock can affect the strength of this muscle. Vitamin C helps.

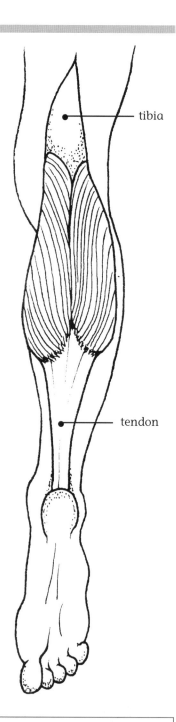

tibia

tendon

WATCH OUT!

Ruptures are common at the point where the Achilles tendon merges with the inner belly of the calf muscle. Usually caused by jumping.

Soleus

(Latin: **soles** - sole, flat).

┌─────────────────────────┐
│ BODY LANGUAGE OF   │
│   WEAKNESS   │
└─────────────────────────┘

Difficulty rising onto toes.

FINGER FIRST AID

Stretch and massage whole muscle. Also press firmly two finger widths above and either side of the navel and either side of the 10th, 11th and 12th thoracic vertebrae.

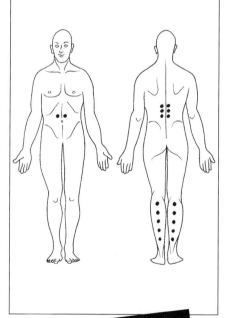

Action

Plantar flexion of foot (in isolation, rising onto demi-pointe with knees bent).

Origin

Outside back of leg, just below the knee.

Insertion

Via the Achilles tendon, into the heel bone.

Synergists

Gastrocnemius, posterior tibial, peroneus longus/brevis, flexor hallucis and digitorum longus.

Alternative connections

Emotional stress, adrenal exhaustion and shock can affect the strength of this muscle. Vitamin C helps.

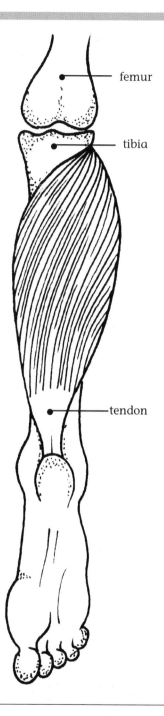

femur

tibia

tendon

WATCH OUT!

Ruptures are common at the point where the Achilles tendon merges with the inner belly of the calf muscle. Usually caused by jumping.

Anterior Tibial

(Latin: **anterior** - in front, **tibia** - shinbone, flute).

BODY LANGUAGE OF WEAKNESS

Weakness in this muscle can lead to 'rolling' inwards due to the collapsing medial longitudinal arch of the foot.

FINGER FIRST AID

Press firmly just above the pubic bone and either side of the 2nd lumbar vertebra. Also massage whole muscle.

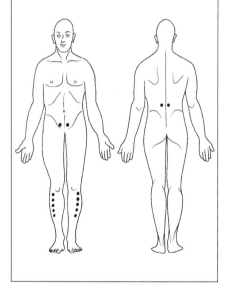

Action

Dorsiflexes (lifts up) and inverts the foot.

Origin

Outer side of tibia, just below the knee at the front.

Insertion

Inner edge of the foot, about two inches from the big toe.

Synergists

Extensor hallucis longus, extensor digitorum longus.

Alternative connections

Associated with bladder problems. Vitamin E helps.

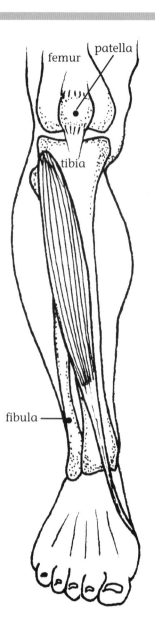

femur patella

tibia

fibula

WATCH OUT!

Often associated with 'shin splints', it is this muscle that aches from jumping on hard surfaces.

Posterior Tibial

(Latin: **posterior** - behind, **tibia** - shinbone, flute).

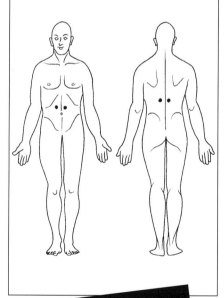
WATCH OUT!

Action

Plantar flexes and inverts the foot. (What every dancer dreads to see when pointing the foot.)

Origin

Back of the tibia and fibula, and the interosseus membrane just below the knee.

Insertion

Medial tarsal and metatarsal bones.

Synergists

Flexor hallucis longus, flexor digitorum longus, gastrocnemius, soleus.

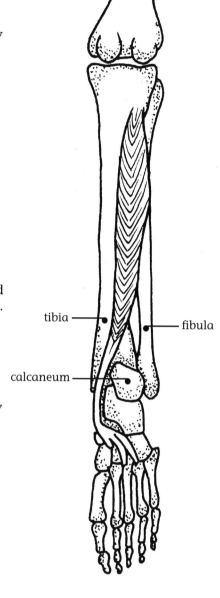

tibia

fibula

calcaneum

Turning out the feet from the ankles rather than the knees causes this muscle to stretch so that the medial longitudinal arch drops.

Peroneus Longus/Brevis

(Greek **perone** - fibula. Latin: longus - long, **brevis** - short).

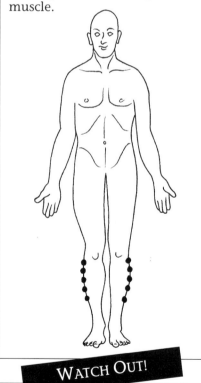

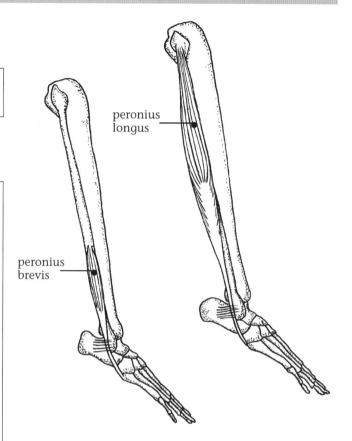

peronius longus

peronius brevis

Action	Plantar flexes (points) and everts the foot. Gives lateral (sideways) stability to the ankle.
Origin	Outer side of the leg, attached to the fibula.
Insertion	Longus: 1st metatarsal. Brevis: 5th metatarsal.
Synergists	Gastrocnemius, soleus.

WATCH OUT!

Spraining this muscle by going over onto the outside of the ankle causes injury which if not treated properly can cause many problems involving the future stability of the joint.

Flexor Hallucis Longus

(Latin: **flexum**- to bend, **hallux** - big toe).

Inability to flex big toe.

FINGER FIRST AID

Massage and stretch under side of foot by pulling the big toe towards the knee.

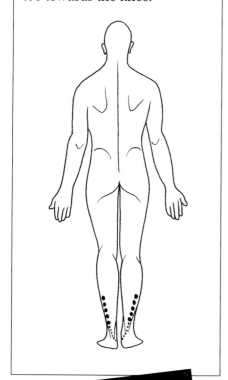

Action

Flexes big toe. Helps stabilise inside of ankle.

Origin

Lower two-thirds of the back of the fibula.

Insertion

Last phalanx of big toe.

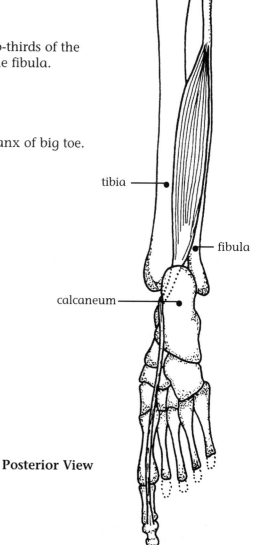

tibia

fibula

calcaneum

Posterior View

WATCH OUT!

The long tendon of this muscle is the weak area and is liable to become inflamed due to excess pressure or overuse.

Extensor Hallucis Longus

(Latin: **extendere** - to stretch, **hallux** - big toe).

Inability to extend big toe.

FINGER FIRST AID

Stretch and massage big toe and top side of foot.

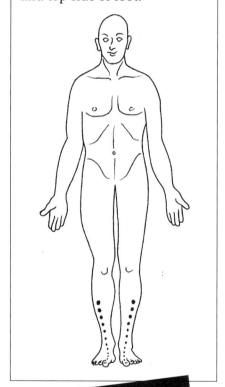

WATCH OUT!

Action

Extends big toe and assists in dorsiflexion of foot.

Origin

Middle half of the front surface of the fibula.

Insertion

Last phalanx of the big toe.

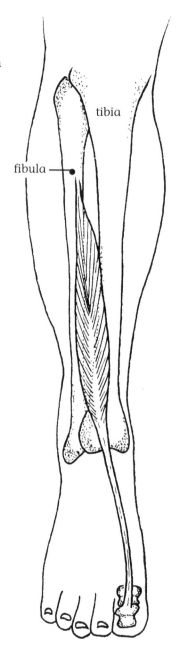

tibia

fibula

The tendon is damaged by compression and by bruising if the toe is stepped on!

Hamstrings

(German: **hamme** - back of leg.
Latin: **stringere** - to draw
together).

'Knock knees', weakness on the
inside of the knee, or 'bow legs'
when the weakness is on the
outside of the knee.

FINGER FIRST AID

Rub origin and insertion of
muscle. Also press into belly of
muscle.

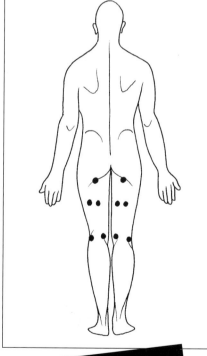

**Posterior
View**

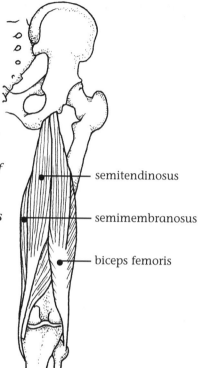

*The hamstrings consist
of three muscles - two
medial (on the inside of
the thigh), called the
semitendinosus and
the **semimembranosus**
and one lateral (on the
outside of the thigh),
called the **biceps
femoris**.*

semitendinosus

semimembranosus

biceps femoris

Action	Flexes the knee, extends the thigh.
Origin	Ischial tuberosity (bottom part of the pelvis) and the lower back portion of the femur.
Insertion	Either side of the tibia, at the back.
Synergists	Gastrocnemius, gracilis, sartorius.
Alternative connections	Problems with these muscles may be associated with toxic conditions of the body involving the large intestine. Also from sacral and pelvic fixations. Vitamin E helps.

WATCH OUT!

These muscles are often injured when dancers do the
splits without proper warm-up. Also when kicking to the front.

Quadriceps

(Latin: **quad** - four, **caput** - head).

*The quadriceps are made up of four muscles. One, the **rectus femoris**, crosses both the hip and knee joint. The other three - **vastus medialis**, **vastus intermedius** and **vastus lateralis** - act on the knee joint only.*

Difficulty straightening the knee and flexing the thigh.

Rub origin and insertion of muscle, the two hollows beneath the patella and generally massage whole length of muscle. Stretch gently.

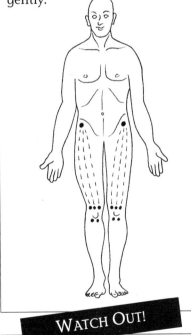

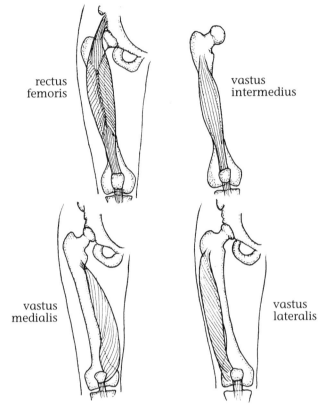

rectus femoris

vastus intermedius

vastus medialis

vastus lateralis

Action	Extends the leg and flexes the thigh.
Origin	Rectus femoris: front point of the ilium. Vastus group; upper portion of the femur.
Insertion	The upper portion of the patella and via the patellar ligament into the tibia.
Synergists	Psoas, sartorius, tensor fascia lata.

WATCH OUT!

Many problems with the knees can be traced back to faulty training techniques which have not taken into account the need to stretch and exercise all parts of the quadriceps. Often dancers suffer from an imbalance between the vastus medialis and the vastus lateralis. This leads to an uneven pull on the patella and is often the precursor to a condition called chondromalacia patella (excess wear on the articular surface between femur and patella).

Sartorius

(Latin: **sartor** - tailor, as in sitting cross-legged).

Weakness of this muscle can cause knee problems, especially on the inside of the knee (knock knees).

FINGER FIRST AID

Massage the whole muscle, paying special attention to the origin and insertion points

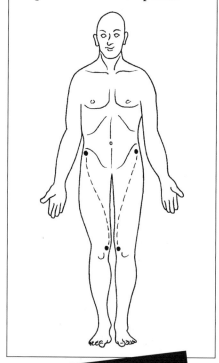

WATCH OUT!

Action

Flexes knee and hip, rotates the thigh laterally (turns outwards).

Origin

Front point of ilium.

Insertion

Medial surface of tibia.

Synergists

Quadriceps, hamstrings, gracilis.

Alternative connections

This muscle is very sensitive to adrenal exhaustion which often manifests as low blood sugar, low resistance to infections and asthma. Vitamin C can help.

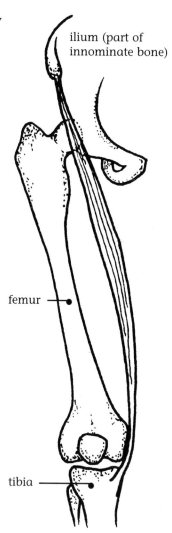

ilium (part of innominate bone)

femur

tibia

Anterior View

Turning out the leg puts great stress on the inside of the knee and often trouble here can be traced to a hypertonic (over-tense) sartorius.

Gracilis

(Latin: **gracilis** - slender).

Lack of pelvic stability, 'knock-knee' problems.

FINGER FIRST AID

Origin and insertion massage, also pressure into the belly of the muscle.

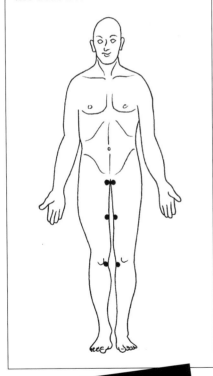

WATCH OUT!

Action

Adducts thigh, flexes knee and hip and medially rotates the thigh and tibia (rotates them inwards).

Origin

Lower edge of pubic bone.

Insertion

Upper part of the inside of the tibia.

Synergists

Adductors, sartorius and hamstring.

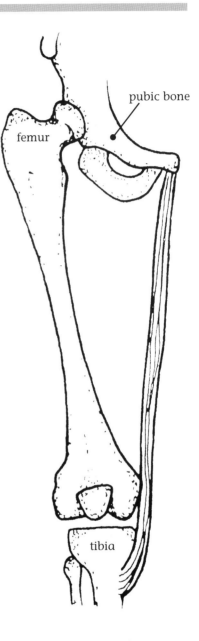

pubic bone

femur

tibia

Anterior View

Groin strains and low back problems involving sacroiliac strains are common. Turning out the legs leads to excess tension in this muscle which is then expected to stretch beyond its limits.

Tensor Fascia Lata

(Latin: **tendere** - to stretch, **fascia** - band, bandage, **lateralis** - side).

Pelvis tilted (up on the weak side). 'Bow legs'. Difficulty standing on one leg with the pelvis level.

FINGER FIRST AID

Massage side of leg from knee to hip. Also the hollow part of the lower back.

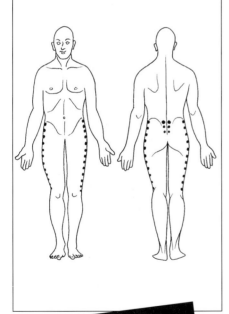

WATCH OUT!

Action

Flexes, abducts and medially rotates the thigh (rotates it inwards). Also stabilises the outside of the knee.

Origin

Outer edge of the ilium, toward the front.

Insertion

Via the long fascia lata tendon to the outside top of the tibia.

Synergists

Gluteus medius and minimus.

Alternative connections

Linked with problems involving the large intestine such as constipation, colitis and diarrhoea. Increase water intake and natural sources of iron.

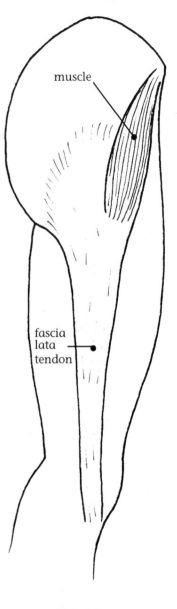

muscle

fascia lata tendon

Side View

Often becomes hypertonic (over-tense), causing lack of pelvic mobility and knee pain on the outside of the knee.

Adductors

(Latin: **ad** - to, **ducere** - to bring).

This group is actually four muscles: the pectineus adductor, the adductor brevis, the adductor longus and the adductor magnus.

Difficulty locking knees and keeping pelvis level in a turned out position.

FINGER FIRST AID

Origin and insertion massage. Also look for hard knots in the belly of the muscle. Press firmly, then stretch slowly.

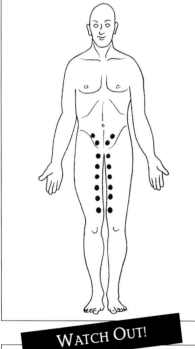

Action

Adduction, hip flexion and lateral rotation of the thigh.

Origin

Front part of pubic bone and lower part of hip bone (ischial tuberosity).

Insertion

Inside of the femur from hip to knee.

Synergists

Gracilis.

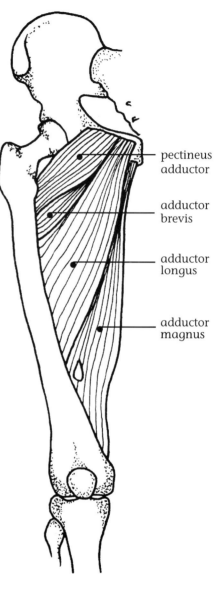

pectineus adductor

adductor brevis

adductor longus

adductor magnus

WATCH OUT!

Groin strains are common with these muscles. They become hypertonic through over-use and when stretched often tear.

Gluteus Medius/Minimus

(Greek: **gloutos** - rump. Latin: **medius** - middle, **minimus** - smallest).

BODY LANGUAGE OF WEAKNESS

Hips uneven, shoulders uneven. Difficulty keeping pelvis level when standing on one leg.

FINGER FIRST AID

Rub firmly: front of body, above the pubic bone and above the thigh; back, inside the prominent knobs on the hip bones at the level of the 5th lumbar joint. Also origin and insertion technique.

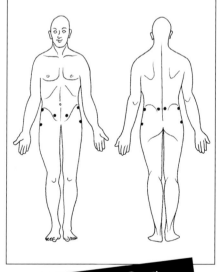

WATCH OUT!

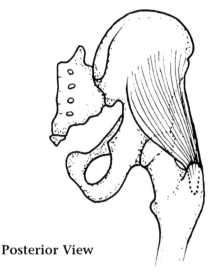

gluteus minimus

This lies under the gluteus medius.

Posterior View

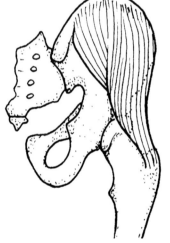

gluteus medius

Action	Abducts thigh, medial rotators.
Origin	Outer surface of ilium.
Insertion	Top of femur to the side.
Synergists	Tensor fascia lata.

Excess tension or weakness in these muscles can lead to problems with the pelvis or knees due to compensation of other muscles.

41

Piriformis

(Greek: **pyramis** - pyramid shaped).

BODY LANGUAGE OF WEAKNESS

Difficulty turning the leg out.

FINGER FIRST AID

Rub firmly the top of pubic bone and above the thigh on the front of body, and inside the prominent knobs on the hip bones at the level of the 5th lumbar joint on the back. Also try deep pressure into the belly (a point between sacrum and femur, beneath the gluteus maximus) - you have to press quite hard, so be careful.

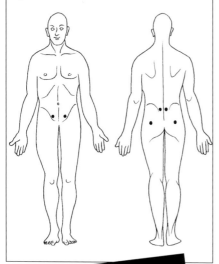

Action

Rotates thigh laterally (turns it out). Abducts thigh when leg is flexed.

Origin

Front of sacrum.

Insertion

Top of femur.

Alternative connections

Problems with the piriformis often affect the sciatic nerve. This can cause pain down the back of the leg, numbness, tingling and sometimes bladder problems.

Posterior View

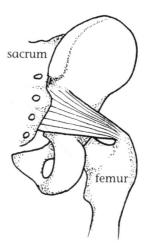

Anterior View

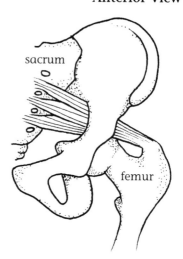

This muscle does most of the work when you turn out the leg.

WATCH OUT!

This muscle usually becomes hypertonic (over-tense), which leads to a restriction of mobility in the hip. Pushing the turn-out without realising the root of the problem can often lead to injury. This is experienced as pain inside the hip or low back, or sciatica (pain down the back of the leg).

Gluteus Maximus

(Greek: **gloutos** - rump. Latin:
maximus - biggest).

BODY LANGUAGE OF WEAKNESS

Excess lumbar curvature, 'sway
back' (pelvis rotated forwards).

FINGER FIRST AID

Origin and insertion technique. Also massage firmly the
vastus lateralis muscle down
the outside of the thigh.

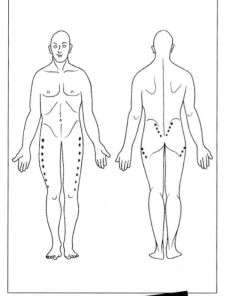

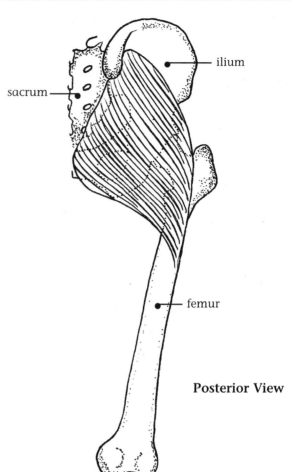

Posterior View

Action	Extends hip, laterally rotates high (assists turn-out).
Origin	Back of the ilium along the sacroiliac joint.
Insertion	Top back of the femur.
Synergists	Hamstrings.

WATCH OUT!

Low back problems due to excess lordosis
(sway back). Difficulty maintaining straight spine when jumping.

Psoas/Iliacus (Iliopsoas)

(Greek: **psoas** - tenderloin. Latin:
ilium - hip bone).

These two muscles will be
considered as one.

Excess or diminished lumbar
curve. Spine tilted away from the
weak side.

FINGER FIRST AID

These muscles are too deep, so
use reflex points. Press deeply
into the abdomen, one inch to
the side and one inch above
the navel. Also on the back,
either side of the 12th
thoracic/1st lumbar joints.

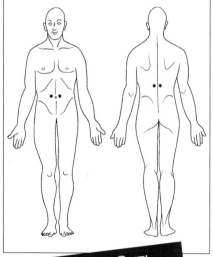

WATCH OUT!

Action

Flexes thigh, aids
slightly in lateral
rotation (turn-
out).

Origin

Psoas: front
surfaces of the
transverse
processes of the
vertebrae from
the 12th thoracic
to the 5th lumbar
joints.
Iliacus: the front
inside of the
ilium and
sacrum.

Insertion

Inside top of the
femur.

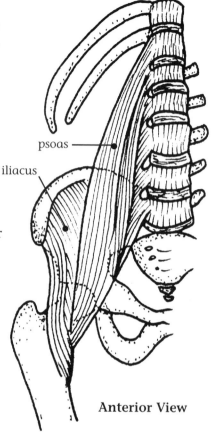

psoas

iliacus

Anterior View

Synergists   Adductors, rectus femoris.

**Alternative
connections**   Nagging low back pain, kidney
disturbances, low energy and foot
difficulties can be associated with
psoas problems. Drink water
regularly. Vitamins A and E help.

This is the main hip flexor and low back stabiliser.
If these muscles go out of balance (usually by
becoming hypertonic), then all kinds of low back and hip problems can develop.

Abdominals

(Latin: **abdomen** - belly)

FINGER FIRST AID

Origin and insertion technique. Also watch the lower part of the rectus abdominals for weakness. If the pelvis is tilted forward when standing, or if there is an excessive hollow in the back when lying down, suspect a weakness here.

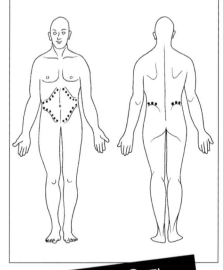

WATCH OUT!

Action

Transverse: constricts abdominal contents, assists in forcing air out of lungs.
Rectus: gives anterior support to lumbar spine, holds rib cage and pubis together.
Internal/external obliques: flex, rotate and side-bend trunk.

There are four main divisions of abdominal muscles

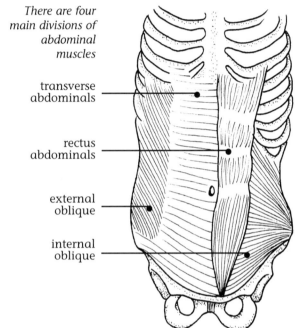

transverse abdominals

rectus abdominals

external oblique

internal oblique

Origin Ribs.

Insertion Upper edges of ilium and pubic bones.

Synergists Iliopsoas.

Alternative connections Digestive problems, particularly those involving the small intestine, are associated with these muscles. Good abdominal tone supports the internal organs, Vitamin E helps.

The lower lumbar joints are most at risk when the abdominals and the iliopsoas muscles have to work together in order to keep the spine stable. One without the other creates an imbalance which eventually will lead to injury.

Bibliography

Anderson, Bob; **Stretching**, Shelter Publications, Bolinas, California, 1987.

Anthony and Thibodeau; **Textbook of Anatomy and Physiology**, The C. V. Mosby Company, St. Louis, Toronto, London, 1983.

Despard, L. L.; **Textbook of Massage**, Oxford University Press, London, 1914.

Hoppenfield, Stanley; **Physical Examination of the Spine and Extremities**, Appleton-Century-Crofts, New York, 1976.

Houston, F. M. and D. C.; **The Healing Benefits of Acupuncture**, Keats Publishing, New Canaan, Connecticut, 1970.

Hurley and Sanders; **Aquarian Age Bio-Engineering**, Los Angeles Health Research, 1970.

Irwin, Yukiko; **Shiatzu**, Routledge & Kegan Paul, London, 1977.

Kapandji, I. A.; **The Physiology of the Joints**, Churchill Livingstone, Edinburgh, London, New York, 1974.

Kapit and Elson; **The Anatomy Colouring Book**, Harper and Row, New York, San Francisco, 1977.

Kennedy, Pat; **The Moving Body**, Faber and Faber, London, 1979.

Kushi, Michio; **How to see your Health: Handbook of Oriental Diagnosis**, Japan Publications Inc., Tokyo, 1975.

McNaught and Callander; **Nurses' Illustrated Physiology**, Churchill Livingstone, Edinburgh, London, New York, 1975.

Pearce, Evelyn; **Anatomy and Physiology for Nurses**, Faber and Faber, London, 1981.

Rowlett, H.; **Basic Anatomy and Physiology**, John Murray, London, 1979.

Sparger, Celia; **Ballet Physique**, Adam and Charles Black, London, 1958.

Thie, John; **Touch for Health**, DeVorss and Company, Marina del Rey, California, 1979.

Topping, Wayne; **Stress Release**, Topping International Institute, Bellingham, 1985.

Walther, David; **Applied Kinesiology**, Systems DC, Pueblo, Colorado, 1981.

Wheeler, C. M.; **Illustrated Human Biology**, Edward Arnold, London, 1978.

Wirhed, Rolf; **Athletic Ability and the Anatomy of Motion**, Wolfe Medical Publications, London, 1984.

Index

Index (cont.)